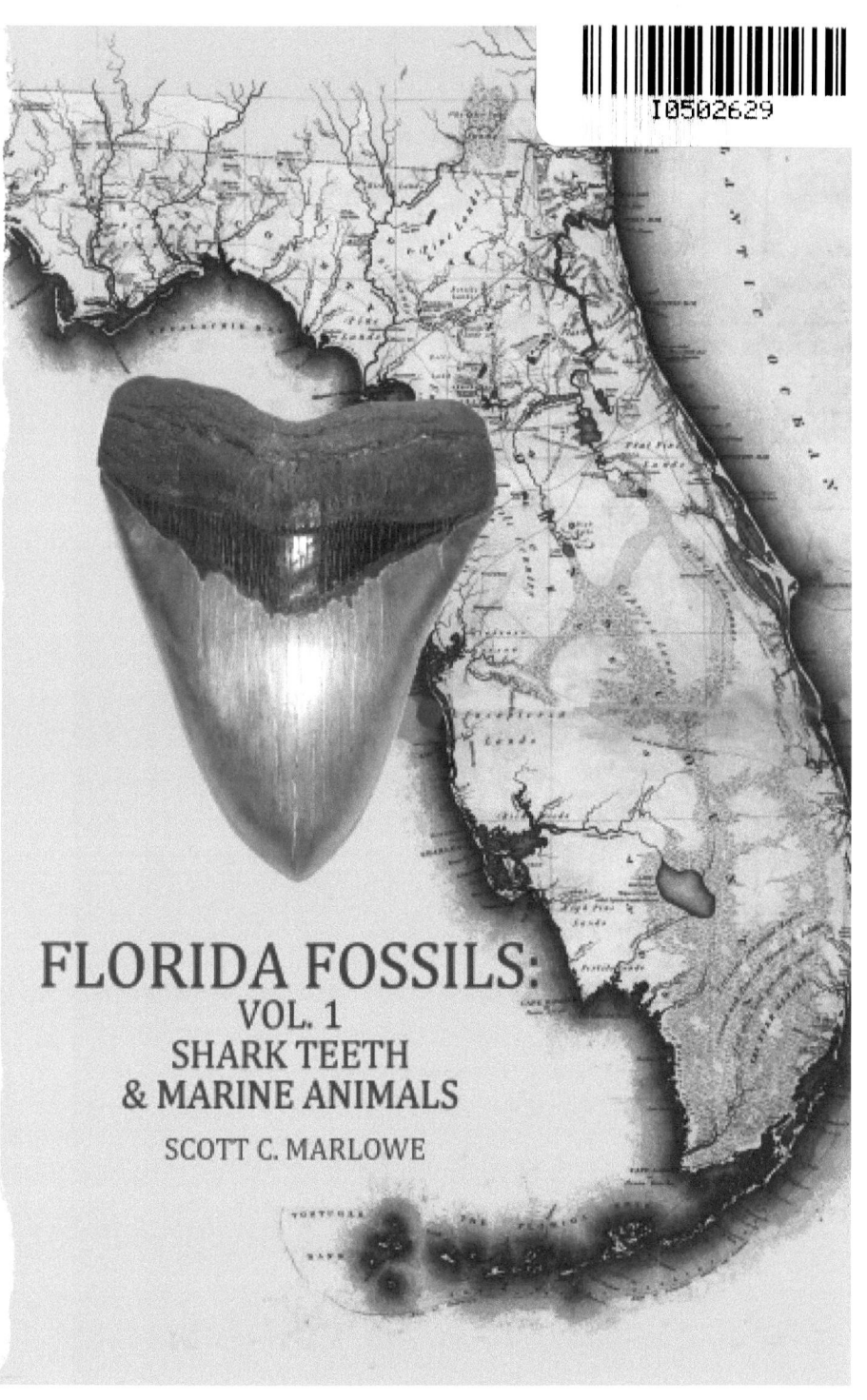

I0502629

FLORIDA FOSSILS:
VOL. 1
SHARK TEETH
& MARINE ANIMALS

SCOTT C. MARLOWE

Cover by Scott Marlowe

First published in the United States by Pangea Press

Pangea Press
514 Winter Terrace
Winter Haven, FL
33881

All rights reserved. No part of this publication may be reproduced stored in any retrieval systems or transmitted in any form, or by any means — mechanical, electric, recording or other — except for rotations utilized in reviews, without the prior authorization and permission of the copyright held author herein.

All authors, artists and media representations cited or provided herein are the copyright of their respective creators. This book is being produced and made available as an educational and scholarly reference resource for and by the author. Wherever possible permission for reproduction herein has been sought, however usage of images and accepted words is covered herein under USC Title 17 Section 107, limitations on exclusive rights. "... for purposes such as criticism, news reporting, teaching . . . scholarship, or research, is not an infringement of copyright."

ISBN: 978-1494881757

FOSSIL HUNTING
ON THE
PEACE RIVER

GEOLOGIC TIME SCALE

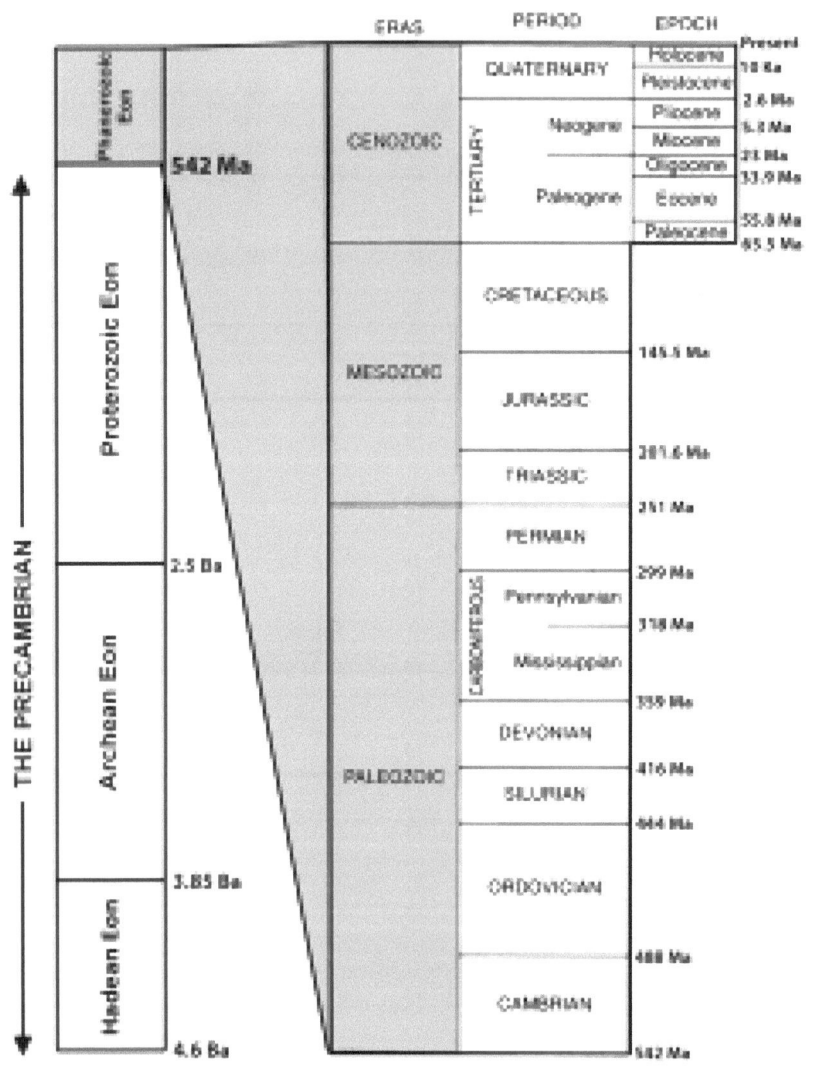

Contents

EXCAVATING A MANATEE SKELETON AT THE VULCAN QUARRY

SHARK'S TEETH

SHARK TOOTH ANATOMY

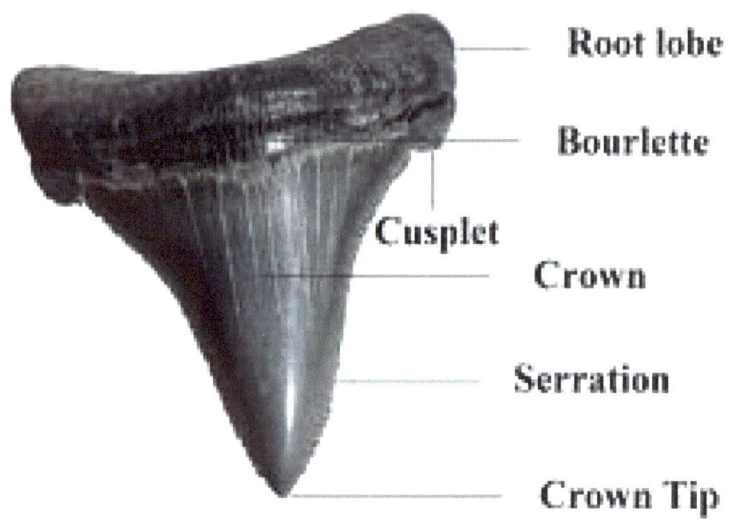

Root lobe

Bourlette

Cusplet

Crown

Serration

Crown Tip

CF Industries Mine Facility

Megalodon Shark (Carcharocles megalodon)
(Extinct). Late Oligocene to early Pleistocene. Found at CF Industries Phosphate Mine, Bowling Green, Florida (Bone Valley).

Great White Shark (Carcharodon carcharias). Miocene to Present. Found in the Peace River near Brownville, Florida.

Mako Shark (Isurus hastalis) (Extinct). Miocene to Pliocene. Found in the Apalachicola River near Alum Cave (Bluff) area, Blountstown, Florida.

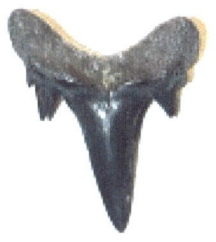

Mackerel Shark (Otodus obliquus) (Extinct).
Paleocene to Lower Eocene. Found in the Suwannee
River near Hart Springs, Fanning Springs, Florida.

Sand Tiger Shark (Carcharias taurus). Miocene
to present. Found at Casperson Beach, Venice
Florida.

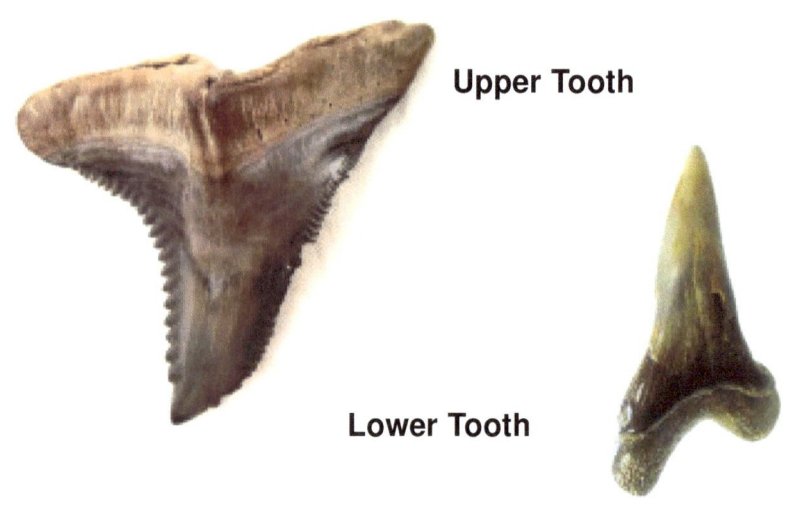

Upper Tooth

Lower Tooth

Snaggletooth Shark (Hemipristis serra) (Extinct). Pliocene to Miocene. Found in the Peace River near Brownville, Florida. Approximately 1 inch in length.

Bull Shark (Carcharhinus leucas). Miocene to Present. Found at Casperson Beach, Venice, Florida. Approximately 1 inch in length.

Tiger Shark (Galicerdo mayumbensis) (Extinct).
Miocene to Plioscene. Found in the Peace River near
Brownville, Florida. Up to 1-1/2 inches in length.

Tiger Shark (Galeocerdo contortus) (Extinct).
Miocene to Pliocene. Found at Casperson Beach,
Venice, Florida. Approximately 1 inch in length.

Lemon Shark (Negaprion eurybathrodon). Miocene to Present. Found at Casperson Beach, Venice, Florida. Approximately 1 inch in length.

Cow Shark (Notorynchus primigenius). Paleocene. Found in the Peace River near Brownville, Florida. From under 1 inch to 1-1/2 inch in length.

Shark Vertebrum (Species Undetermined).
Miocene to Pliocene. Found in the Peace River near
Brownville, Florida. Diameter varies.

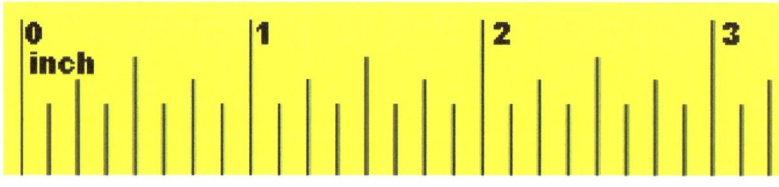

Sting Ray Barb and Teeth (Probably Myliobatis sp.)
Eocene to Present. Found in the Peace River near
Brownville, Florida. Lengths vary.

**OTHER
MARINE
FOSSILS**

Sea Biscuit Echinoid (Rhyncholampas gouldii).
Oligocene. Found at the Vulcan Quarry near Brooksville, Florida. Sizes vary.

Sea Biscuit Echinoid (Rhyncholampas evergladesensis). Pliocene to Pleistocene. Found at the Vulcan Quarry near Brooksville, Florida. Sizes Vary

Sand Dollar (Durhamella ocalana). Eocene. Found at the Vulcan Quarry near Brooksville, Florida. Sizes Vary.

Sea Urchin (Gagaria mossomi). Oligocene. Found at the Vulcan Quarry near Brooksville, Florida. Sizes vary.

INSPECTING A WASH POOL FOR FOSSILS AT VULCAN QUARRY

Baracuda Tooth
(Sphyraena barracuda)
Miocene
Found in the Peace River
near Nocatee Bridge,
Nocatee, Florida.
About 1-1/2 inch in length.

Alligator Gar Fish Tooth
(Atractosteus spatula)
Miocene to Pleistocene
Found in the Peace River
near Arcadia, Florida
About 1-1/2 inch long.

Sawfish Tine
(Pristis sp.)
Oligocene
Found in the Peace River
near Nocatee Bridge,
Nocatee, Florida.
Up to 2-1/2 inches in length.

Puffer Fish Mouth Plate
(Chilomycterus sp.)
Miocene to Pleistocene
Found in the Peace
River near Arcadia,
Florida.
Sizes Vary

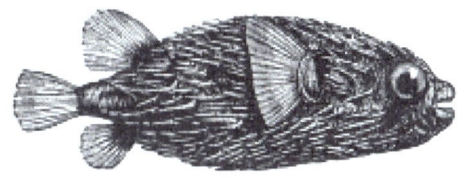

PUFFER (PORCUPINE) FISH

Drum Fossil Fish Teeth
(Pogonias sp.)
Miocene to Pleistocene
Found in the Peace
River near Brownville,
Florida.
About 1/4" in Diameter.

Dolphin tooth (Hadrodelphis sp.) Miocene. Found in the Peace River near Brownville, Florida. Approximately 1-1/2 inches in length.

Opposite Page:

Dolphin Inner Ear (Periotic Bone) (Hadrodelphis sp.) Miocene. Found in the Peace River. Location unknown. About 1 inch in size.

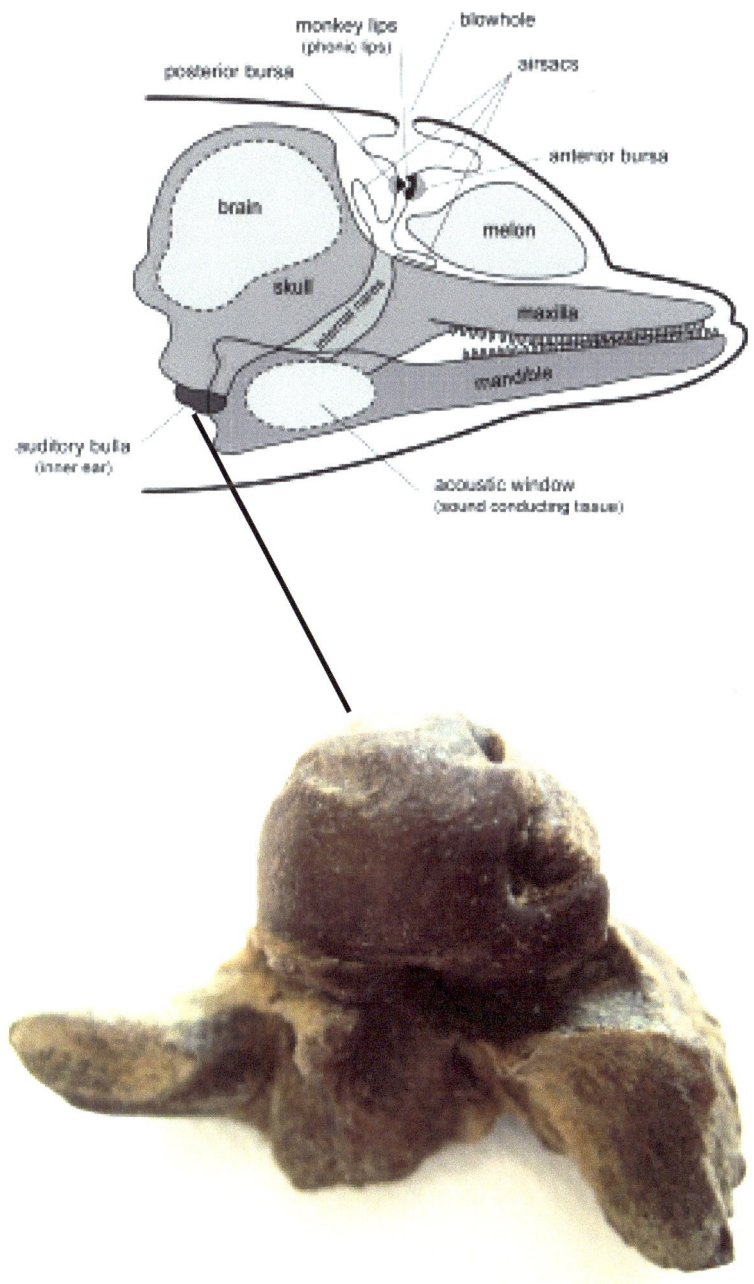

Diagram labels: posterior bursa, monkey lips (phonic lips), blowhole, airsacs, anterior bursa, brain, melon, skull, internal nares, maxilla, mandible, auditory bulla (inner ear), acoustic window (sound conducting tissue)

Whale Vertebrum (Probably Balaenoptera_). Miocene. Found in Bradenton, Florida during a dig at a private property location. Approximately 9 inches in height.

Sperm Whale Tooth (Physeter sp.). Pleistocene, Found at CF Industries Quarry near Fort Meade, Florida. About 6 inches in length.

Dugong Rib Section (Metaxytherium floridanum). Miocene. Foun in the Peace River near Brownville, Florida. Sizes Vary.

Alligator Tooth (Alligator mississippiensis).
Pleistocene
Found in the Peace River
near Brownville, Florida
About 1-1/2 inches long.

Alligator Osteoderm (Scute) (Alligator mississippiensis).
Pleistocene
Found in the Peace River
near Brownville, Florida
About 1-1/2 inch Diameter

Crocodile Tooth Embeded in Jaw Bone (Gavialosuchus amercanus)
Miocene
Found in High Springs, Near
Gainesville, Florida
Sizes Vary

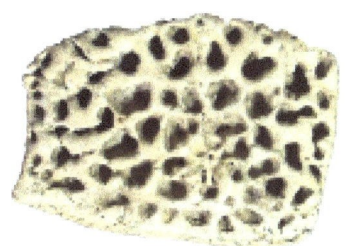

Crocodile Osteoderm (Scute) (Gavialosuchus amercanus)
Miocene
Found in High Springs, Near
Gainesville, Florida
Sizes Vary

HUNTING FOSSILS
ON THE PEACE RIVER
NEAR BROWNVILLE, FLORIDA

FOSSIL RESOURCES

Brevard Museum of History &
Natural Science
2201 Michigan Ave.
Cocoa, FL 32926
312/632-1830

Bailey-Matthews Shell Museum
3075 Sanibel-Captiva Road
Sanibel, FL 33957
239/395-2233

Camp Bayou Outdoor Learning
Center
4140 24th St SE
Ruskin, FL 33570
813/641-8545

Clewiston Museum
109 Central Avenue
Clewston, FL 33440
863/983-2870

Dinosaur World
5145 Harvey Tew Road
Plant City, FL 33565
813/717-9865

Discovery Museum
2701 W. Cypress Creek Road
Fort Lauderdale, FL 33309
800/882-0278

Florida Museum of Nat. History
3215 Hull Road
Gainesville, FL 32611
352/846-2000

Mulberry Phosphate Museum
101 SE 1st St
Mulberry, FL 33860
863/425-2823

Museum of Arts and Sciences
352 S. Nova Road
Daytona, FL 32114
386/255-0285

Museum of Seminole County
History
300 Bush Boulevard
Sanford, FL 32773
407/665-2489

Shell Factory
2787 N Tamiami Trail
Fort Myers, FL 33903
239/995-2141

Shell World
99801 Overseas Hwy
Key Largo, FL 33037
305/451-9797

Shell World
4727 W Irlo Bronson Mem'l Hwy
Kissimmee, FL 34746
407/396-2242

South Florida Museum
201 10th St W
Bradenton, FL 34205
941/746-4131

The Dinosaur Store
250 W. Cocoa Beach Cswy.
Cocoa Beach, FL 32931
321-783-7300

Windley Key Fossil Reef Geo-
logical State Park
U.S. 1 Mile Marker 84.9
Islamorada, FL 33036
305/664-2540